ISBN: 979-8-9947729-0-4

Published by RGBinder, LLC

Printed in the United States of America

Dedicated to my family, especially Caliber, who delights in "science experiments" involving color, and Emeri, who loves to sing about color wherever she goes. Their curiosity, creativity, and joy in discovering how the world looks helped inspire me to write this book.

And now I'd like to share that same sense of discovery with you.

I've spent my career working in printing and color management, helping pressrooms and brands understand, measure, and control color. This book is a simpler, kid-friendly version of what I have taught for years: how light, materials, and our eyes work together to create the colors we think we see.

Welcome to your journey with the Color Detectives and their sidekick, Light Beam. Light Beam is a playful little beam of light who appears throughout the book to point out clues and help you understand how light, materials, and your eyes work together.

For activity and illusion support, plus new content, visit our Color Detectives website: www.rgbinder.com/colordetectives

Questions or feedback? Email colordetectives@rgbinder.com.

The following items would be helpful for the cases:

- 2 or 3 Flashlights (phone flashlights will work)
- Colored transparent film for use as flashlight filters (red, green, and blue are perfect)
- White paper
- Magnifying glass (basic handheld is fine)
- Pencil and notepaper
- CD, Prism or "suncatcher" crystal for refraction (rainbow) experiments
- In lieu of the above a transparent glass with water will work
- A few everyday objects with different surfaces (glossy package, matte paper, fabric, aluminum foil, colored wrapper) for comparing how materials affect color
- Colored paper and scissors
- Adult help for cutting/handling small parts and for any lamp or light setup

This book belongs to:

____

____

____

The Rainbow Inside Light

The Color Detectives examine the beam from their flashlight. The light is...just white.

Are we sure though?

Let's investigate!

Case #1

The flashlight looks white, but the rainbow shows it is made of many colors.

Wow!

The light has colors inside it!

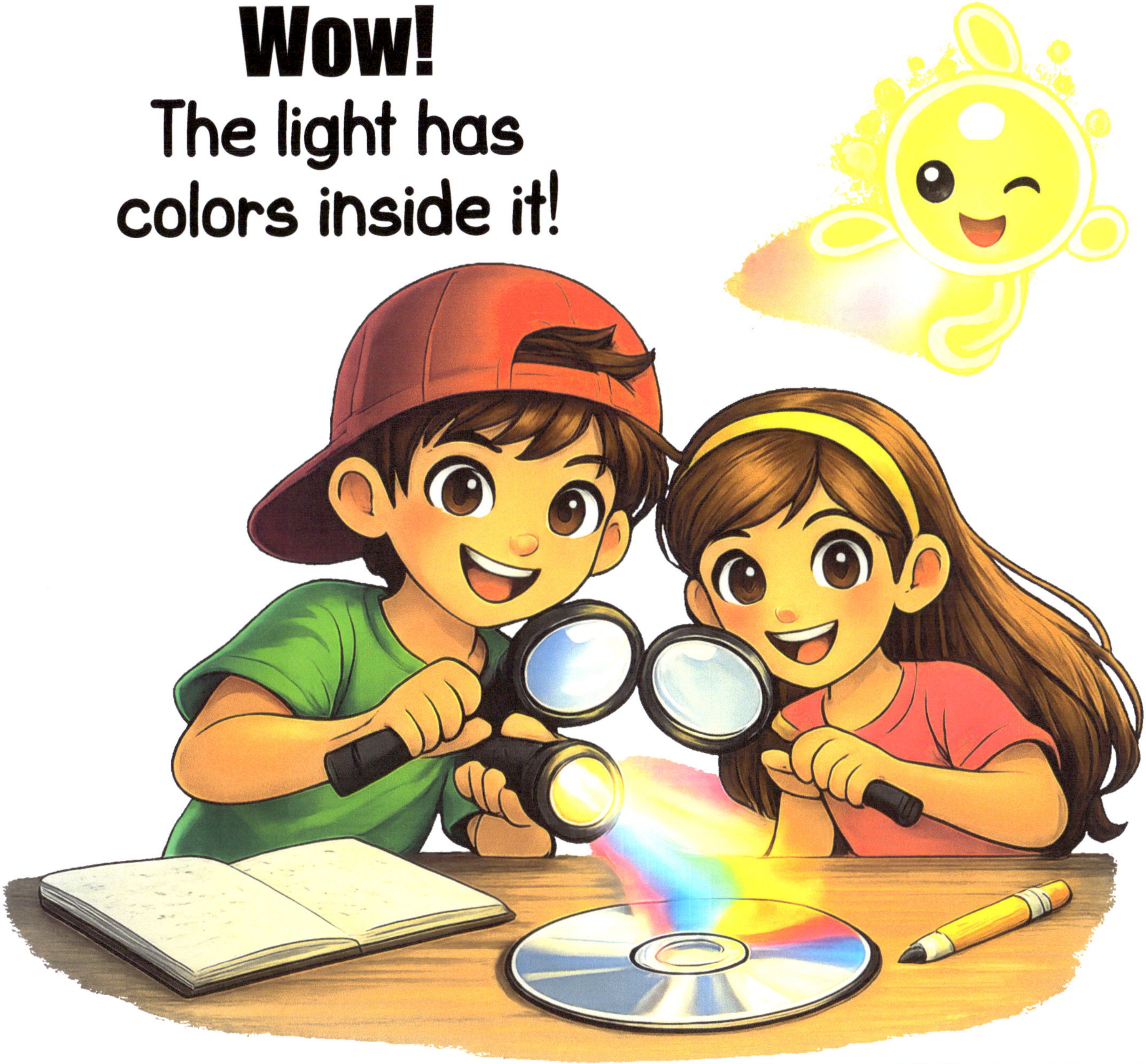

Make a Rainbow from "White" Light

1. Get a flashlight and a shiny CD.
2. Shine the flashlight on the CD.
3. Tilt it until you see a rainbow!

Reflection and Absorption

The Color Detectives want to know how our eyes and brain work together to see all the colors around us.

The color we see comes from reflected light.

Light Beam points out that objects reflect some colors while absorbing other colors!

Light shines all around us reflecting off all the objects we see.

Our eyes detect the reflected light with special cells called cones.

Your brain gets signals from your eyes, then interprets the colors!

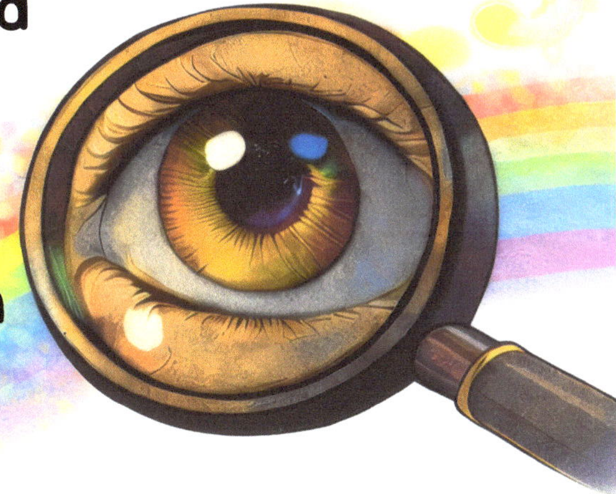

1. Collect some different objects.
2. Look at how they reflect or absorb light.
3. Which colors do you see?
   Which colors are missing?

Lighting and Color

The Color Detectives got an up close look at how different lights affect color.

Let's investigate!

Case #3

Different lights can change the way that colors look.

1. Shine different lights on objects.
2. Carefully watch how the color changes.
3. Try using different white or colored lights.

Color Illusions

The Color Detectives wondered if their two oranges are the same color.

Let's Investigate!

Our eyes and brains work together but illusions can fool us!

1. Look closely at the illusions on the following pages.
2. What are your eyes and brain seeing?
3. Try to understand why you are seeing what you do.

After staring at the red heart, then looking at the pure white surface, you will see the opposite color.

What Color do you see?

There is no right or wrong answer because we all see color a bit differently from one another.

In the illustration below, which circle is darker, A or B?

A

B

Both circles are the same shade of gray!

The surrounding colors make the circles look different.

1. Cut multiple shapes from colored paper.
2. Lay them over different background colors.
3. Can you make the colors look different?

Mixing Colors

The Color Detectives tried mixing colors with paint and light. Light Beam appeared. Let's investigate!

Paint and light mix to make new colors, but they do it differently.

Mixing paint or ink is subtractive. Adding more paint or ink creates new darker colors.

Mixing light is additive. Adding more light creates new lighter colors.

1. Try mixing different paint colors using a brush. Start with light colors and mix them to make darker colors. Write down which colors you mixed and what new colors you created.

2. Ask an adult to help you find translucent film to use as filters for flashlights. You will need two or more flashlights. Overlap the beams of light to make new colors. Write down which colors you overlapped and what new colors you saw.

3. Look very briefly at a phone screen, computer monitor, or television using a magnifying glass. Can you see tiny red, green, and blue dots blending together to make the images you see?

Printing and Dots

The Color Detectives got a close-up look at printed pictures, and saw an array of dots.

Let's Investigate!

Printing uses tiny dots to make pictures!

When printers make pictures, they don't paint smooth colors like a brush does.

Instead, they use tiny dots of color placed very close together.

Your eyes mix these tiny dots to make new colors.

From far away, the dots look like one smooth picture even though they are really many small dots!

This is how books, posters, and newspapers make colorful images using ink.

1. Use a magnifying glass to look closely at a printed picture.
2. Can you see the tiny dots?
3. How do the dots mix to make the colors?

Let's Review

The Color Detectives, along with their friend Light Beam have learned a lot about color and how we see it. Let's review.

Color happens when light, your eyes, and your brain work together.

Light

All of the colors that we can see come from the light.

Eyes

Our eyes take light that reflects off of the objects around us sending signals to our brain.

Brain

Our brains process the information from our eyes and we see color.

For Parents, Teachers, and Educators

The cases in this book are designed to introduce foundational ideas about color through observation, discussion, and hands-on exploration. They are intentionally presented without technical vocabulary so children can focus on noticing patterns, asking questions, and testing ideas for themselves.

The pages that follow provide background explanations for each case, written for adults. They describe what is happening scientifically, explain why the activity works, and offer optional extension activities suitable for classroom or home use. These notes are intended to support instruction, guide discussion, and help answer common "why" questions that arise during exploration.

This section is not required for reading the book with children. It is included as a resource for educators and caregivers who want additional context or wish to extend the investigations beyond the story pages.

Big Ideas in This Book

This is not just another typical color book for children. In the included cases, children learn that:

- Light is the source of all color
- Objects change how light behaves
- Different lighting changes how colors look
- Our eyes and brains work together to create what we see
- Color mixing depends on whether you are mixing light (additive) or paint/ink (subtractive)
- Printed pictures are made from tiny dots that our eyes blend together

These ideas form the foundation for later learning in art, science, printing, photography, and digital media.

How to Use This Book

Read the story pages aloud and pause to look closely at the pictures.

Encourage children to describe what they notice before explaining anything.

Use the "Try It!" sections as invitations, not requirements. Even imagining the activity is valuable.

There are no wrong answers—differences in observation are part of the lesson.

About the Illusions

Some pages include visual effects that may surprise children (and adults). This is intentional. The message is simple: Our eyes can be fooled, and that's normal. Also, we all see color differently so we may not all have the same experience with the illusions.

For Educators

This book aligns naturally with:

- Early science inquiry
- Visual arts
- STEAM discussions
- Media literacy (how pictures are made)

The activities are designed to be low-cost, flexible, and adaptable to classroom or home settings.

Final Thought

Color is not just something we see; it is something we perceive.

This book invites children to begin that journey with curiosity, creativity, and confidence.

Appendix

Parent & Teacher Guide: What's Happening Behind Each Case

This appendix provides background explanations and optional extension activities for adults supporting children through the Color Detective cases. The child-facing pages intentionally avoid technical language. The explanations below are for adult understanding only and are not required for children to enjoy or benefit from the book.

Case 1: Light Contains Many Colors

White light is made up of many different wavelengths of light energy. When white light passes through or reflects off certain objects—like a CD—it separates into a rainbow. This happens because different wavelengths bend and refract to reveal the component colors.

The CD works like a simple diffraction grating, spreading the light out so the colors become visible.

Why It Matters

This establishes the most important idea in the book:

- Color begins with light
- Objects do not create color on their own; they change light

Optional Extension Activities

1. Use a prism or glass of water in sunlight to make a rainbow.
2. Compare rainbows made with sunlight vs. indoor lighting.

Ask: Does every white light make the same rainbow?

Case 2: Objects Reflect Some Colors and Absorb Others

Objects appear a certain color because they reflect some wavelengths of light and absorb others.

A red apple primarily reflects red wavelengths while absorbing most others.

Why It Matters

This explains why color depends on both:

- The light source
- The object being viewed
- It prepares children to understand why the same object can look different in different lighting.

Optional Extension Activities

1. Shine colored light on different objects and observe changes. Cool white vs. warm white bulbs, or flashlights with colored film as filters work great.
2. Compare light-colored vs. dark-colored materials.

The changes will be subtle, but careful observation will reveal differences.

Case 3: Lighting Changes What Colors Look Like

Different white lights are not identical.

Some contain more red energy, others more yellow or blue. These differences subtly change how objects appear—even when the object itself has not changed.

This is why color matching can be difficult across rooms or times of day.

Why It Matters

- Children learn that color is not fixed
- Lighting is part of what we see

Optional Extension Activities

1. Compare an object under daylight, warm indoor light, and cool LED light.
2. Take photos of the same object in different lighting and compare.
3. Ask: Which looks most "real"? Why might people disagree?

Case 4: Eyes and Brains Work Together

This case includes two classic perception effects.

Retinal Fatigue (Afterimage Effect)

When we stare at a strong color for a period of time, the color-sensitive cells in the eye (cones) responding to that color become temporarily less sensitive.

When the child looks away at a white surface, the brain interprets the remaining signals as the opposite color.

Why It Matters

It shows that:

- Seeing is an active process
- The brain interprets signals, it doesn't just record them

Extension Activity

1. Try different colors and record which opposite color appears.
2. Ask children if everyone sees the same result.

Simultaneous Contrast

A color's appearance is influenced by surrounding colors.

The brain exaggerates differences to improve contrast, which can make identical colors appear different.

Why It Matters

This explains why:

- Color judgments depend on context
- Backgrounds matter in art, design, and printing

Extension Activity

1. Place the same gray shape on different colored papers.
2. Ask children to evaluate how the color appears to change.

Case 5: Mixing Light Is Different from Mixing Paint

Light mixing is additive:

- Red, green, and blue light combine at different intensities to create many different colors
- 100% red, green, and blue light create white

Printing ink is subtractive:

- Layers of transparent yellow, magenta, and cyan ink combine to make darker colors by filtering specific wavelengths of light
- 100% yellow, magenta, and cyan ink create black

Paint is opaque, and a range of various pigments are blended to control which specific wavelengths of light energy are reflected or absorbed to create different specific colors.

Why It Matters

It is often assumed that all mixing works the same way. This case separates systems they will encounter in display screens, art, and printing.

Optional Extension Activities

1. Mix paint colors to observe how blending different paint creates new colors. Are the new colors darker or lighter than the original paints?
2. Use overlapping colored flashlights in a dark room - red, green, and blue if possible. What new colors are formed when the colors overlap?

Use a magnifying glass to see the red, green, and blue pixels that create images on a tablet, phone, or monitor.

Case 6: Printing Uses Tiny Dots

Printed images are made from tiny dots of color. From a distance, the eye blends them together. Up close, the dots are visible.

This relies on how human vision averages small details.

Why It Matters

This introduces the concept of how images are built, which ties directly to how books, packaging, and other printed materials that surrond us every day are created.

Optional Extension Activities

1. Look at magazines or printed packages with a magnifying glass.
2. Compare printed dots to pixels on a screen.
3. Ask: Why do we see a picture instead of dots?

Case 7: Seeing Colors Everywhere

This case reinforces that color perception is:

- Based on light
- Influenced by surroundings
- Interpreted by the brain
- Experienced differently by different people

Why It Matters

It encourages curiosity instead of certainty. Children learn that noticing differences is part of learning.

Optional Extension Activities

1. Ask children to find color "mysteries" in daily life.
2. Encourage drawing or photographing color changes.
3. Discuss how artists, scientists, and printers all work with the same color rules.

Final Note for Adults

You do not need to explain everything.

The most important lesson children learn from this book is that looking closely matters, and that questions are more valuable than answers.

For activity and illusion support, plus new content, visit our Color Detectives website: www.rgbinder.com/colordetectives

Questions or feedback? Email colordetectives@rgbinder.com.

www.ingramcontent.com/pod-product-compliance
Ingram Content Group UK Ltd.
Pitfield, Milton Keynes, MK11 3LW, UK
UKHW060107300726
14090UKWH00003B/400

9798994772904